John Joseph Griffin

On the Reaction of Ethyl and Methyl Alcohol With

Paradiazometatoluenesulphonic Acid

John Joseph Griffin

On the Reaction of Ethyl and Methyl Alcohol With Paradiazometatoluenesulphonic Acid

ISBN/EAN: 9783744669764

Printed in Europe, USA, Canada, Australia, Japan

Cover: Foto ©berggeist007 / pixelio.de

More available books at **www.hansebooks.com**

ON THE

REACTION OF ETHYL AND METHYL ALCOHOLS

WITH

PARADIAZOMETATOLUENESULPHONIC ACID

IN THE

PRESENCE OF CERTAIN SUBSTANCES;

AND ON

METATOLUENESULPHONIC ACID.

DISSERTATION

SUBMITTED TO THE BOARD OF UNIVERSITY STUDIES

OF THE JOHNS HOPKINS UNIVERSITY, FOR THE

DEGREE OF DOCTOR OF PHILOSOPHY,

BY

JOHN J. GRIFFIN.

1895.

ACKNOWLEDGMENT.

The work herein described, was undertaken at the suggestion of Professor Ira Remsen, and pursued throughout under his guidance. I hereby express my gratitude to him for his kind counsel and instruction during my course at Johns Hopkins University.

I also express my obligations to Professor Morse, of the Department of Chemistry, and to Professor Ames and Dr. Hulburt, under whom I pursued the studies of physics and mathematics respectively.

CONTENTS.

THE REACTION OF ETHYL AND METHYL ALCOHOLS WITH PARA-DIAZO-META-TOLUENE-SULPHONIC ACID IN PRESENCE OF CERTAIN SUBSTANCES.

                                                                    Page.

Introduction 1
Preparation of Para-diazo-meta-toluene-sulphonic acid. 2
Decomposition of the same in Methyl Alcohol in Presence
    of Sodium Methylate 5
In Presence of Sodium Carbonate. 8
In Presence of Sodium Hydroxide. 11
In Presence of Zinc Dust 12
In Presence of Calcium Carbonate 13
Decomposition in Ethyl Alcohol in Presence of Sodium
    Ethylate. 15
In Presence of Sodium Carbonate. 18

	Page
In Presence of Sodium Hydroxide	20
In Presence of Zinc Dust	21
In Presence of Ammonia	22
Comparative Yields of Amide by Various Methods of Decomposition	24
Experiments with Phosphorus Trichloride	26
Meta-toluene-sulphon-amide	31
Analysis of Meta-toluene-sulphon-amide	35
Decomposition of same	38
Oxidation of same	39
Meta-sulphamine-benzoic Acid	40
Barium Salt	41
Silver Salt	43

META-TOLUENE-SULPHONIC ACID.

Introduction	46
Preparation of Meta-toluene-sulphonic Acid	59
Barium Salt	63
Lead Salt	64
Calcium Salt	66

	Page
Sodium Salt.	
Potassium Salt	69
Silver Salt.	70
Magnesium Salt	71
Manganese Salt	72
Zinc Salt.	73
Copper Salt.	75
Preparation of Meta-toluene-sulphon-anilide.	76
The Toluides of Meta-toluene-sulphonic Acid	76
Conclusion	78
Biographical Sketch.	81

-----------oOo-----------

ON THE REACTION OF ETHYL AND METHYL ALCOHOLS WITH PARA-DIAZO-META-TOLUENE-SULPHONIC ACID.

INTRODUCTION.

Since it became known that the statement of Griess to the effect that, when diazo compounds are boiled with alcohol, the decomposition products are nitrogen, aldehyde, and a substance formed by the replacement of the diazo group by hydrogen, was not at all general; but that in some cases, the ethoxy group, and not hydrogen, enters into the compound, various efforts have been made to determine the conditions that decide which of these two reactions, known as the alkoxy and hydrogen reactions respectively, will take place. Work was done in this laboratory on the influence of pressure and temperature on the reaction, and it was shown that increase of pressure and simplicity of the alcohol radical favor the alkoxy reaction.

Metcalf,[1] in studying the reaction of certain alcohols with para-diazo-meta-toluene-sulphonic acid, showed that, when the diazo compound is decomposed in alcohol, both the alkoxy and hydrogen reactions take place, the former greatly predominating, though the ratio changes somewhat with varying conditions of temperature and pressure. A few years later, Beeson,[2] in his investigations on the action of certain diazo compounds on methyl and ethyl alcohols, carried on in this laboratory, having discovered the fact that the presence of an excess of alkalies and of some other substances determined the hydrogen reaction, it was thought desirable to study similar reactions in the case of para-diazo-meta-toluene-sulphonic acid.

PREPARATION OF PARA-DIAZO-META-TOLUENE-SULPHONIC ACID.

As a starting point, the para-toluidine-meta-sulphonic acid of commerce was taken. This was a light brown powder, slightly soluble in cold, readily soluble in boiling water. After boiling with animal charcoal, two, or at the most,

[1] American Chem. Journal, 15 - 301; [2] Ibid 16 - 235.

three, recrystallizations gave a pure white product, separating from the solution in prismatic needles. By adding a little hydrochloric acid with the animal charcoal to the boiling solution, the first crystallization yielded a nearly white product. 100 grams of the crude material yielded 85 grams of the pure acid.

Pechmann[1], who obtained para-toluidine-meta-sulphonic acid by treating para-toluidine with fuming sulphuric acid, describes the product as of a sulphur yellow color which it was impossible to remove, and Metcalf, who followed the same method of preparation, speaks of the difficulty of obtaining the acid in a pure white condition.

The pure white para-diazo-meta-toluene-sulphonic acid thus obtained, was finely pulverized, and diazotized by suspending it in alcohol of 92° and treating it with a strong current of nitrous fumes, produced by the action of nitric acid on arsenious oxide, in the manner described by Metcalf[2], and with substantially the same results.

100 grams of the acid were suspended in 30 cc. of al-

[1] Liebig's Annalen, 173 — 195.
[2] Dissertation, 1890, and American Chemical Jour. 15 — 301.

cohol, and the nitrous fumes passed through the emulsion for thirty minutes. The action was very slow at first, but soon increased as the mixture warmed up. The alcohol took on an orange color which deepened to a dark brownish red shade. The diazo compound was of a light pink tint, and settled readily to the bottom of the flask after shaking, while, on the other hand, the alcohol with the suspended acid would remain milky for some time. This rapid settling to the bottom was taken as indicative of the completion of the reaction. The diazo compound was filtered off by means of a filter pump, washed with a small quantity of alcohol, and some ether, and then dried. The yield in this and subsequent experiments, approached closely to the theoretical amount, never falling below 95%. Some of the diazo compound dissolved in the alcohol, and on standing, crystallized out. The compound when dry, keeps for months in the dark, but exposed to the light, it slowly turns yellow. It is very stable, and small portions have been ground in a mortar without exploding.

DECOMPOSITION OF PARA-DIAZO-META-TOLUENE-SULPHONIC ACID IN METHYL ALCOHOL, IN PRESENCE OF VARIOUS SUBSTANCES.

I. *In presence of sodium methylate.* - In 80 cc. of pure methyl alcohol, redistilled over lime, 5 grams of metallic sodium were dissolved. The solution of sodium methylate was then placed in a beaker in a freezing mixture, and, when its temperature fell below zero, the para-diazo-meta-toluene-sulphonic acid was added, small portions at a time. There was immediate action, a gentle evolution of gas took place, the diazo compound gradually disappeared, and the alcohol took on a straw yellow color. After the addition of a few small portions, the temperature rose to 5° and the color deepened. No further addition of the diazo compound was made until the thermometer had again fallen below zero, and the temperature was kept below that point throughout the experiment. When about half the diazo compound had been added, an orange-yellow powder began to settle in the bottom of the beaker. The addition of the diazo compound was continued until 25 grams had been added; the beaker was

then removed from the freezing-mixture and left to stand over night. During the decomposition, a slightly irritating odor, other than that of the alcohol, could be observed. When the reaction had ceased, the material was poured into an evaporating-dish, and the methyl alcohol driven off by heating on a water-bath. The substance readily dried. When nearly dry, some volatile product, having a very irritating odor, and acting upon the eyes to produce an immediate flow of tears, began to be given off. The hard, light-brown substance which remained in the dish, was pulverized, and rubbed in a mortar, with its own weight, (30 grams) of phosphorus pentachloride. The reaction was immediate and violent, and fumes of phosphorus oxychloride were given off. There remained behind a dark brown oil, together with a quantity of tarry products. This was poured into water, and a dark-brown oil settled in the bottom of the vessel. Placed for some time in a test-tube, in a freezing-mixture, the oil did not solidify. This oil was washed with water several times and finally treated with strong ammonia, with which it readily reacted, going into solution, and heating up considerably. Excess of ammonia was expelled by evapor-

ating to dryness on a water-bath, and the residue extracted several times with ether. Upon distilling off the ether, there remained behind a brown, syrupy liquid which on cooling became solid and showed a crystalline structure. This was dissolved in boiling water and treated with animal charcoal, and filtered. Upon cooling, small rhomboidal scales separated out, slightly brown in color. By several recrystallizations and treatment with animal charcoal, large clear white fern-shaped crystals were obtained, which melted at 108°. This was the melting-point of the small quantity of meta-toluene-sulphon-amide obtained by Metcalf from a mixture consisting for the greater part of para-ethoxy-meta-toluene-sulphon amide. The yield was 4.5 grams.

Another experiment was tried under similar conditions, with the exception of that of temperature, which was allowed to rise to 20°, and the results were essentially the same, with perhaps a slight increase in the tarry by-product. In this case, instead of pouring the mixture resulting from the decomposition of the diazo compound into an evaporating-dish, and driving off the alcohol, it was placed in an Erlenmeyer flask, and distilled off. There remained behind

a light brown colored, porous mass, easily removed from the flask. This was heated on a water-bath till thoroughly dry. The same irritating odor was noticed. The alcohol which was distilled off was tested for aldehyde and it responded at once to the silver mirror test. A solution of the light-brown powder was also tested for aldehyde, and its presence was indicated in the same way. By treatment with phosphorus pentachloride and ammonia, as in the first case, four grams of the amide were obtained. In these and the subsequent experiments, no product other than the meta-toluene-sulphon-amide was obtained. Its formation must, therefore, have been through the following steps:

$$C_6H_3\begin{cases}CH_3\\SO_3H\,(m)\\N_2\,(p)\end{cases} \longrightarrow C_6H_3\begin{cases}CH_3\\SO_3\,(m)\\N_2\,(p)\end{cases} \longrightarrow C_6H_4\begin{cases}CH_3\\SO_3H\end{cases} \longrightarrow$$

$$C_6H_4\begin{cases}CH_3\\SO_3Na\end{cases} \longrightarrow C_6H_4\begin{cases}CH_3\\SO_2Cl\end{cases} \longrightarrow C_6H_4\begin{cases}CH_3\\SO_2NH_2\end{cases}$$

II. In the presence of sodium carbonate. - 25 grams of the diazo acid were decomposed in 200 cc. of absolute methyl alcohol, in which 9 grams of dried sodium carbonate were suspended. When the alcohol had been cooled below zero,

the addition of small portions of the diazo compound was begun. As the amount of substance added increased, the alcohol gradually took on a pink tinge, but, beyond an occasional bubble at long intervals, there was no apparent evolution of gas. The flask was then removed from the freezing-mixture, placed on a water-bath, and gently warmed to 30°, when a slow evolution of gas occurred. The flask was left standing over night. In the morning the alcohol was found to be colored brownish red, and the sodium carbonate formed a layer at the bottom, seemingly in undiminished quantity. A thin layer of undecomposed or undissolved diazo compound covered the sodium carbonate. An occasional bubble of gas would rise through this upper layer. It was placed on a water-bath and heated to 35°, when the action became quite rapid, and continued till nearly all the sodium carbonate had disappeared. The temperature was then 46°. The flask was next connected with a condenser, and nearly all the alcohol was distilled off. The distillate gave a decided response to the test for aldehyde. The residual brown powder was very easily moved from the flask. Dried on the water-bath, it weighed 27 grams, and was darker in

color than that obtained in the experiments with sodium methylate. 35 grams of phosphorus pentachloride were placed in a 500 cc. Erlenmeyer flask; the 27 grams of dark brown powder then added, and the flask connected with a condenser in order to obtain the volatile products of the reaction. The action began in less than a minute, and the contents of the flask became quite hot, though the action was not nearly as violent as in the experiment with sodium methylate. A great quantity of oxychloride fumes came over, but only a few drops of oxychloride were condensed in the receiver. The contents of the flask were poured into a large quantity of water, and a heavy, dark oil collected at the bottom of the vessel. This oil, after thorough washing, was poured into a quantity of concentrated ammonia, and well shaken. After a few moments, the contents of the flask heated up gently, and the oil disappeared, the ammonia turning a brown color. Left to stand over night, some brown, feathery crystals, together with some drops of a dark, gummy substance, covered the bottom of the flask. The contents of the flask were placed in an evaporating-dish, and evaporated to dryness on a water-bath. The dry material was extracted

several times with ether, the ether distilled off, and a brown, crystalline substance obtained. Boiled with water and animal charcoal, a yield of 12 grams of a light-brown colored amide was obtained. A second experiment with sodium carbonate and methyl alcohol was performed, the alcohol being at the room temperature, 20°. A gentle evolution of gas began at once, and the temperature rose gradually while the diazo compound was added. It was not allowed to rise above 45°, being cooled when it reached that point by immersion in cold water. The remainder of the process was exactly as in the previous case, resulting in a like yield, 12 grams of amide.

III. <u>In the presence of sodium hydroxide</u>. - In this experiment, 7 grams of fused sodium hydroxide were broken up and placed in an Erlenmeyer flask with 200 cc. of methyl alcohol. Small portions of the diazo compound were added, and the action was immediate, the temperature, which was 21° at the beginning of the experiment, rising so rapidly that the flask was placed in ice-water and kept there throughout the experiment. After 25 grams of the diazo compound had

been added, the flask was left standing till all action had ceased, when the alcohol was distilled off as in the preceding cases. On applying the silver mirror test to the alcohol which was distilled off, aldehyde was found to be present in it. After the alcohol had been evaporated off, and the residual gummy mass dried and powdered, it was treated with phosphorus pentachloride. The action was extremely rapid and violent. The thick black oil which resulted, was washed with water, and then treated with strong ammonia, resulting in a yield of 12 grams of amide.

IV. In the presence of zinc dust. - In a flask containing 200 cc. of absolute methyl alcohol, 10 grams of zinc dust were placed. The diazo compound was then added in small portions, the flask being well shaken after every addition. Action took place at once with a rapid rise of temperature from $17°$, at the beginning, to $30°$. When this latter temperature was reached, the flask was placed in ice-water. In this experiment the alcohol did not discolor for a long time, whereas in the other decomposition, a very small proportion of the whole quantity of the diazo com-

pound sufficed to give the alcohol a decided red color. When 25 grams of diazo acid had been added, the alcohol was of a light straw-color.

After all action had ceased, the solution was filtered to remove the residual zinc dust. The alcohol was then poured into a flask, connected with a condenser and distilled. A yellowish, syrupy liquid remained behind. This was poured into an evaporating-dish, and heated on a water-bath to dryness. During the drying, the irritating odors of the previous experiments were noticed. When dry, the substance was ground up in a mortar, and 30 grams of an almost white powder were obtained. Treated with phosphorus pentachloride, the action was very violent, great heat was developed, and but a small quantity of oil was obtained. This, however, was almost white, not like the dark viscid oil of the previous experiments. Treated with ammonia in the usual manner, 5 grams of pure white amide, melting at $108°$, were obtained.

V. <u>In presence of calcium carbonate</u>. - In this experiment 300 cc. of methyl alcohol and 15 grams of powdered

calcium carbonate were taken. The diazo compound was added
in small portions at a time; but no action was noticed.
The flask was then placed on a water-bath, and heated up to
45°, but still there was no evolution of gas. The alcohol,
however, became pink, probably from some of the diazo compound going into solution. 50 grams of the diazo acid were
added, and the whole left to stand over night. In the morning, the calcium carbonate covered the bottom of the flask,
with the lighter, undissolved diazo compound forming a layer above.

From this experiment, it seems, therefore, that the
presence of calcium carbonate has no influence whatever on
the diazo decomposition; which entirely agrees with Beeson's[1]
results in his experiments with diazo-benzene-nitrate and
calcite.

An excess, 17 grams, of zinc dust, was then added to
the mixture. Bubbles of gas formed at once, and as the contents of the flask heated up, the reaction went on briskly.
The alcohol, which was of a pink color, soon became colorless, and at the conclusion of the action, possessed the

---
[1] American Chemical Journal, 15 - 235; and Dissertation.

light yellow color observed in previous experiments with zinc dust. After filtering from the calcium carbonate and residual zinc dust, drying, and proceeding as before, a yield of 10 grams of pure white amide was obtained.

DECOMPOSITION OF PARA-DIAZO-META-TOLUENE-SULPHONIC ACID IN ETHYL ALCOHOL IN PRESENCE OF VARIOUS SUBSTANCES.

I. *In presence of sodium ethylate.*— 3.5 grams of sodium were dissolved in 20 cc. of absolute ethyl alcohol, and the solution placed in a 500 cc. Erlenmeyer flask and cooled down to $-5°$. The diazo compound was added in small portions at a time. Each portion sank at once to the bottom of the flask, but in a few moments rose rapidly to the surface, colored a dark red, and an evolution of gas began. The odor of aldehyde was very marked from the beginning. As the addition proceeded, the solution thickened, and became very dark in color. The evolution of gas became sluggish, and endured for a long time after all the diazo compound had been added. The temperature was kept below zero

until 25 grams of the diazo compound had been added, when the flask was removed from the freezing-mixture, and the contents left to attain the room temperature.

The mixture was allowed to stand over night, then placed upon a water-bath, and the alcohol distilled off. A hard black mass remained in the flask. This could not be removed, and the alcohol was poured back into the flask and heated on the water-bath; but the mass remained hard. It was left standing two and a half days, but with no better result. The alcohol was distilled off again, and the mass dissolved readily in a small quantity of water. This solution was evaporated to dryness over a water-bath. The drying was a long and tedious process, the mass becoming gummy. This same difficulty in drying the product which had been in contact with water, was noticed by Metcalf.[1] The dried mass was powdered and weighed 45 grams. It was of a dark brown color, much darker than the corresponding compound obtained from methyl alcohol. This was placed in a 500 cc. Erlenmeyer flask, which, after the addition of 45 grams of phosphorus pentachloride, was rapidly connected with a con-

---
[1] Dissertation, page 6; and Amer. Chem. Journal, 15 -30].

denser, by means of rubber tubing, and shaken. The reaction took place very slowly, the flask heated up gradually, and a few drops of phosphorus oxychloride distilled over. There remained a black viscous mass. This was shaken up with ether; the ether solution poured off, and the ether distilled off.

A thick black oil remained, which was shaken up several times with water. Placed in a freezing-mixture for half an hour, it did not solidify. This oil was then shaken up with ammonia, but the action was very feeble, and after the action some tar remained. The ammonia solution was poured off, evaporated to dryness, and the residue extracted several times with ether. 4 grams of dark-colored amide were obtained. The experiment was repeated with the same quantities of material, but no efforts were made to keep the temperature down. The only difference noted was in the rapidity of the reaction which, in the second experiment, was much greater than when the flask was kept in the freezing-mixture, but in neither case was it nearly as brisk as when methyl alcohol was employed. A yield of 3 grams of amide and a greater quantity of tarry matter than in the previous

experiment, were obtained. The alcohol which was distilled off in both these cases, gave the mirror test for aldehyde.

II. *In the presence of sodium carbonate.*—9 grams of dried sodium carbonate were suspended in 200 cc. of absolute alcohol, and, after cooling the mixture to zero, 25 grams of the diazo compound were gradually added. There was no apparent action, beyond the alcohol becoming colored a pale yellow. Minute bubbles appeared after a few minutes. As the action seemed not to increase, the flask was taken from the freezing-mixture and placed on a water-bath. When its temperature was $32°$, the action became brisk, but not nearly as much so as in the case of methyl alcohol and sodium carbonate. The odor of aldehyde was distinctly perceptible. The flask was kept on the water-bath till its contents attained a temperature of $50°$. It was then removed, and left standing over night.

In the morning the alcohol was colored a deep red, and bubbles of gas were still rising to the surface. Upon heating the flask, the evolution of gas became very rapid. When this had almost entirely ceased, the alcohol was distilled

off. The brown mass remaining behind was removed very easily, and powdered. It weighed 35 grams. This was rubbed in a mortar with an equal weight of phosphorus pentachloride. The action was gentle, and the product was shaken up with water, and then extracted with ether. 26 grams of a dark brown oil, which did not thicken in a freezing-mixture, were obtained. This was poured into 100 cc. of concentrated ammonia in an Erlenmeyer flask, and the flask and contents were shaken. The oil disappeared in a few minutes, the ammonia heated up and became brown in color. After evaporating to dryness, extracting with ether, and recrystallizing from water and animal charcoal, 10 grams of a pure white amide were obtained.

A second experiment was performed under similar conditions, with the exception that the oil obtained by the pentachloride treatment, was not extracted with ether, but, after washing with water, poured into the concentrated ammonia.

The amide obtained after recrystallizing from water and animal charcoal weighed 8 grams, and was of a light brown color.

A third experiment was performed, the variation in this

case consisting in using 90% alcohol in place of absolute alcohol. This experiment was performed at the room temperature. Upon the addition of the diazo compound, there was a brisk evolution of gas but no odor of aldehyde. The contents of the flask heated up to above 40°. After standing some time, it was heated to 45°, when the bubbles became larger, and the effervescence proceeded briskly for some time, the sodium carbonate gradually disappearing till but a small portion of it remained. The odor of aldehyde was now perceptible. From this it seems that the reaction took place in two stages, the first being probably the formation of the sodium salt of para-diazo-meta-toluene-sulphonic acid, and the second the decomposition of this compound.

Proceeding as in the second experiment, 7 grams of a light-brown amide were obtained.

III. In presence of sodium hydroxide. — In this experiment, an excess, 7 grams, of fused sodium hydroxide was broken up into small pieces and placed in an Erlenmeyer flask with 200 cc. of absolute ethyl alcohol. The experiment was begun at the room temperature. The reaction was

similar to that with sodium ethylate. The decomposition took place much more slowly than when ethyl alcohol and sodium hydroxide were used. After distilling off the alcohol, drying and powdering, 35 grams of a brown substance were obtained. Treating with phosphorus pentachloride, a black oil resulted. Upon washing this with water, quite a quantity of black tarry scum rose to the surface. The yield of amide obtained from this oil was 10 grams, and its color was brown.

IV. In the presence of zinc dust. — In this experiment, 300 cc. of absolute ethyl alcohol and 17 grams of zinc dust were taken. 50 grams of the diazo compound were slowly added. No freezing-mixture was employed, but the temperature was kept down by placing the the flask in running water. The evolution of gas was a little sluggish at first, but, as the contents of the flask heated up, it became brisk. The odor of aldehyde was noticed from the very beginning. When the reaction was over, the alcohol remained limpid and of a light orange color, in distinction to the dark-red, syrupy liquid which resulted from the other experiments with

ethyl alcohol. The alcohol was distilled off, leaving a hard, almost white residue on the bottom of the flask. This was removed by boiling with water, and the watery substance was then evaporated to dryness, which was readily done, and powdered, yielding 55 grams. Treated with phosphorus pentachloride, the action was very violent, and a small quantity of a brown oil was obtained. This, after washing with water, gave 7 grams of a nearly pure white amide.

V. *In presence of ammonia.* — 300 cc. of absolute methyl alcohol were saturated with dry ammonia gas, and the solution was cooled to zero. Upon the addition of the diazo compound, there was no apparent action. After waiting some time, the flask was removed from the freezing-mixture and its contents were allowed to take the room temperature (20°), when action began at once. As more diazo compound was added, the temperature rose rapidly, and the flask was cooled by placing it in a dish of cold water. Upon the completion of the reaction, the alcohol was distilled off, the residue removed from the flask, dried and powdered. The powder was treated in a mortar with an equal amount of phos-

phorus pentachloride; the reaction was very violent; fumes of phosphorus oxychloride were given off, but in place of the appearance of an oil, the mixture in the mortar thickened and became solid. Water was added to this, upon which the mass foamed up, and after the reaction was over, bunches of yellow needles appeared throughout the dark red liquid.

A quantity of crystals was removed, and washed with water. They were found to be very soluble in water. Repeated boiling with animal charcoal failed to remove or lessen their yellow color. Heated up to 300°, they charred without melting. Their solution gave a neutral reaction. On adding strong hydrochloric acid to a solution of the substance in a test-tube, white needles, similar in appearance to the original para-toluidine-meta-sulphonic acid crystallized out. It was suspected that it might be the ammonium salt of para-toluidine-meta-sulphonic acid, and to test this, about 20 grams of it was diazotized in the usual manner, and the diazo compound, similar in appearance to that obtained heretofore, was then decomposed in absolute methyl alcohol to which an excess of fused sodium hydroxide had been added. The decomposition took place in all re-

spects similar to that of the diazo acid in methyl alcohol and sodium hydroxide. On drying, powdering and treating with phosphorus pentachloride, a dark-colored oil was obtained, which was washed and treated with ammonia. The resulting amide was extracted with ether, boiled with water and animal charcoal, and brown crystals were obtained which, by repeated purifications, gave large pure white scales, of exactly the same appearance as the recrystallized amides previously obtained, and melting at the same point.

The experiment was repeated, using ethyl alcohol, with like results.

Hence the decomposition of para-diazo-meta-toluene-sulphonic acid in ammoniacal alcohol, must have taken place as follows:

$$C_6H_3{\begin{matrix}CH_3\\SO_3\\N_2\end{matrix}} \longrightarrow C_6H_3{\begin{matrix}CH_3\\SO_3H_4N\\NH_2\end{matrix}} \longrightarrow$$

$$C_6H_3{\begin{matrix}CH_3\\SO_2Cl\\NH_2\end{matrix}} \longrightarrow C_6H_3{\begin{matrix}CH_3\\SO_2OH\\NH_2\end{matrix}}$$

A comparison of the different yields of purified amides was made as follows, the numbers representing the yield in grams of pure amide obtained from the decomposition of 100

grams of the para-diazo-meta-toluene-sulphonic acid.

	In methyl alcohol.	In ethyl alcohol.
With sodium	24.8	19.
" sodium carbonate	33	33.
" sodium hydroxide	33	24.
" zinc dust	17	24.

Each weight is the average of at least three experiments.

From these experiments, it is seen that the sodium carbonate reaction yielded the largest amount of amide. The compound resulting from the decomposition of the diazo acid in presence of sodium carbonate, and which contained the excess of sodium carbonate originally added, was also the one most slowly acted upon by phosphorus pentachloride.

On the other hand, with the exception of the zinc dust reaction, all the others gave a quantity of tarry matter upon treatment with phosphorus pentachloride.
The apparent purity of the oil obtained by treating with phosphorus pentachloride the product of the decomposition of para-diazo-meta-toluene-sulphonic acid in the presence of zinc dust; the absence of all traces of tarry matter, and

the large clear white crystals of amide which the very first crystallization yielded, marked the zinc reaction as by far the cleanest method of decomposition. The great disparity in the yields of amide obtained by this and the other methods of decomposition, was not, however, in its favor.

It was thought that the heat of the reaction might have decomposed or volatilized a part of the sulphon-chloride formed, and experiments were made with a less violent chlorinating agent than phosphorus pentachloride. Accordingly 50 grams of the diazo compound were decomposed in 300 cc. of methyl alcohol and 17 grams of zinc dust. After the decomposition, the solution was filtered and evaporated to dryness, 55 grams of light-brown powder being obtained. This was added in small portions at a time, to 70 grams of phosphorus trichloride, contained in a 300 cc. round bottomed flask. After a few grams had been added, small bubbles began to appear, coming from the powder, which darkened somewhat in color, and became gummy. As the addition of substance went on, the temperature rose very slightly. After the whole had been added, the flask was put aside over night.

In the morning, the contents of the flask consisted of a brown gum which covered the bottom, and an almost clear liquid on top. Most of the clear liquid, the greater part of which was phosphorus trichloride was poured off, placed in a flask and distilled under diminished pressure. There remained behind a small quantity of a slightly yellow oil. This was shaken up with water; it became lighter in color, diminished somewhat in quantity, and settled readily to the bottom of the flask.

The flask containing the gummy substance and the remainder of the clear liquid, was then connected with a condenser, and heated upon a water bath under diminished pressure. Some more phosphorus trichloride came over. When distillation had ceased, the flask was removed, and the slightly yellow liquid above the gummy material poured into cold water. It settled at once to the bottom. Water was added to the flask containing the brown gum, and this dissolved slowly, giving off small globules of oil which gathered together at the bottom. The three portions of oil were put together, and washed repeatedly with water.

The amide was made by pouring the oil into a 500 cc.

Erlenmeyer flask containing from 150 to 200 cc. of strong ammonia, and shaking vigorously. In a few minutes the contents of the flask became warm and the oil dissolved forming a clear, light-yellow solution. Upon cooling, large slightly yellow crystals of the amide separated out. After filtering, the mother-liquid was evaporated to dryness and the residue extracted with ether, which resulted in a still smaller addition to the total amount of amide obtained. After boiling with animal charcoal, the first crystallization yielded from 100 grams of the diazo compound, 45 grams of large pure white fern-shaped crystals, which melted sharply at $105°$.

The decomposition was repeated, using absolute ethyl alcohol in place of methyl alcohol. The decomposition did not take place as readily as when methyl alcohol was employed, but, upon slightly warming the alcohol, it became rapid. The alcohol became nearly colorless throughout the experiment, while when methyl alcohol was used, it turned to an orange color. The yield of amide was a little greater than by the methyl alcohol method; 100 grams of the diazo acid giving 50 grams of pure amide.

The result obtained by heating the product of the diazo decomposition in presence of zinc dust with phosphorus trichloride, suggested a like experiment with that obtained by decomposing the diazo acid in presence of sodium carbonate.

Because the amides obtained by treating the residue, after the decomposition in presence of sodium carbonate, with phosphorus and ammonia respectively approached that from the zinc dust reaction in whiteness and surpassed it in quantity, it was thought the substitution of the trichloride for the pentachloride of phosphorus, would lead to a larger and cleaner yield of meta-toluene-sulphonchloride.

Accordingly, about 25 grams of the diazo compound were decomposed in absolute ethyl alcohol in which 8 grams of dried sodium carbonate were suspended.

The reaction took place as in the like experiments previously performed, and about 30 grams of dried material were obtained. This was powdered, and added, little by little to 30 grams of phosphorus trichloride contained in a round-bottomed flask. Each small portion added settled to the bottom, and there was no appearance of any change

taking place. When the whole amount was added, the powder absorbed the liquid trichloride, but the contents of the flask did not warm up, nor was any other indication of a reaction noticed. Left to stand over night, its appearance was unchanged in the morning.

The flask was next connected with a reflux condenser, and heated on a water bath for two hours. A portion of the moist powder was then removed from the flask and thrown into cold water. There was the usual reaction between phosphorus trichloride and water, but no trace of an oily sulphon-chloride. The substance was, therefore, removed from the flask and rubbed in a mortar with 30 grams of phosphorus pentachloride, resulting in the usual quantity of liquid meta-toluene-sulphon-chloride, which was converted into the amide.

From this experiment it was concluded that the product obtained when the diazo compound was decomposed in presence of sodium carbonate would not react with phosphorus trichloride and all subsequent yields of sulphon chloride were obtained by the zinc dust and phosphorus trichloride treatment.

META-TOLUENE-SULPHON-AMIDE.

Having on hand a quantity of pure amide, obtained as a result of the preceding experiments, its properties were carefully studied, and compared with the meta-toluene-sulphon-amides described in the literature.

The amide was first mentioned, in 1873, by Müller,[1] who made it by reducing the barium salt of ortho-brom-meta-toluene-sulphonic acid by means of sodium amalgam; and then by the phosphorus pentachloride reaction forming the sulphon-chloride, which, treated with ammonia, gave the amide. The transition from the ortho-brom-meta-toluene-sulphonic acid to the amide, he supposed took place as follows:

$$C_6H_3\begin{cases}CH_3\\Br\,(o)\\SO_3H\,(m)\end{cases} \longrightarrow C_6H_4\begin{cases}CH_3\\SO_3H\end{cases} \longrightarrow$$

$$C_6H_4\begin{cases}CH_3\\SO_2Cl\end{cases} \longrightarrow C_6H_4\begin{cases}CH_3\\SO_2NH_2\end{cases}$$

Müller describes the amide as easily soluble in ether, alcohol and ammonia, and as crystallizing from water at 40°

---

[1] Liebig's Annalen, Vol. 169 - p.61.

in large scales, and from water at $20°$ in slender needles. Its melting-point was $90° - 91°$.

A year later, Pechmann[1] obtained what he claimed to be meta-toluene-sulphon-amide by decomposing para-diazo-meta-toluene-sulphonic acid in alcohol under pressure, according to the following equation:

$$C_6H_3 \begin{subarray}{c} CH_3 \\ SO_3(m) \\ N_2 \end{subarray} + C_2H_5OH = C_6H_4 \begin{subarray}{c} CH_3 \\ SO_3H \end{subarray} + C_2H_4O + N_2$$

From this acid he obtained an amide, like Müller's soluble in hot water, alcohol and ether, but which crystallized in rhombic scales and in brilliant leaves and needles, and melted "somewhere below $100°$."

The amide was next obtained in 1875, by Pagel[2], who also employed the diazo reaction, making first the diazo compound of ortho-toluene-sulphonic acid, changing this into the brom-toluene-sulphonic acid, removing the bromine by long-continued treatment with sodium amalgam, and then proceeding to the amide by the usual reactions. Pagel's amide was tolerably soluble in cold water, much more so in hot water, and easily soluble in alcohol, chloroform and ether.

---

[1] Liebig's Annalen, 173 - 195. [2] Ibid, 176 - 297.

From all these solvents it crystallized out in large leaves which melted at 104°.

In 1877, Beckurts[1] described an amide which he obtained by fractional crystallization from a mixture of amides. But this product, and that obtained two years later by Claesson and Wallin[2] was a mixture of the para and ortho compounds, as will be shown in the second part of this thesis.

In 1879, F.H.S.Muller,[3] by treating the meta-diazo-toluene salts with sulphurous acid, obtained an amide melting at 107°. This also was probably impure, for an anilide obtained from the same sulphon-chloride, had a much lower melting-point than the one which will be described further on.

Neville and Winther,[4] (1880), by heating ortho-diazo-toluene-sulphonic acid under pressure with alcohol, and by reduction of bromo-toluene-sulphonic acid by means of sodium amalgam, obtained a toluene-sulphonic acid, which gave, on treating its potassium salt with phosphorus pentachloride, a sulphon-chloride, and an amide which melted at

---

[1] Berichte X — 913.   [2] Ibid XII — 1848.   [3] Ibid VII — 1416.
[4] Journal of the Chemical Society, Vol. 37, p. 623.

$106.5°- 107.5°$

Chase Palmer,[1] in 1882, repeated Lubner and Müller's process of obtaining the amide, and found a product identical in all its properties with that described by these investigators.

In 1883, Klason[2] prepared the amide after the method of Pechmann, and found it to have a melting-point of $108°$, crystallizing in monoclinic scales and octahedra.

In this same year, Vallin,[3] admitting the evidence of the impurity of the amide obtained by himself and Klason,[4] made the compound by the method of Pechmann, and described it as crystallizing from water in leaves melting at $107°$, and from alcohol in monoclinic crystals which melted at $108°$.* About this time, also, Noyes and Walker[5] made what they claimed to be the meta-toluene-sulphon-amide by treating ortho-brom-toluene with fuming sulphuric acid, making

---

[1] American Chemical Journal, 4 — 140. [2] Berichte 19 — 2587.
[3] Ibid 19 — 2952. [4] Ibid 19 — 1845. [5] American Chemical Journal, 6 — 165.

* The names of these two investigators were written Claesson and Wallin, in Berichte XII, -- 1848.

the sodium salt and eliminating the bromine atom by means
of zinc dust. The sodium salt was then treated with phosphorus pentachloride and the oily chloride thus obtained
converted into an amide which melted at 90°— 91°.
The next contribution to our knowledge of meta-toluene-sulphon-amide was made by Metcalf,[1] in his work "On the Reaction
of Certain Alcohols with Para-diazo-meta-toluene-sulphonic
acid", which showed that when this acid is decomposed by
boiling with alcohol, both the alkoxy and the hydrogen reactions take place, the former to a much greater degree than
the latter. Metcalf succeeded in separating a small quantity of the meta-toluene-sulphon-amide from mixtures of it
and para-ethoxy-meta-toluene-sulphon-amide. He found for
these specimens, melting-points ranging between 95°— 108°.

    The amide which was obtained in the experiments here-
in described, was carefully compared with that described by
the above mentioned investigators. From concentrated solutions, the amide came down in brilliant hexagonal scales,
and from more dilute solutions it separated out in beauti-
ful large fern-like growths. It was soluble to a very
slight extent in cold water, but easily soluble in hot water

---
[1] Dissertation, 1890, and American Chem. Journal 15 - 301.

at a temperature of 70°. It was exceedingly soluble in alcohol, from which it crystallized in large monoclinic prisms.

Analyses of the amide gave the following results:

I. .3020 gram of the amide gave .5415 gram of carbon dioxide.

II. .3003 gram of the amide gave .5425 gram of carbon dioxide.

III. .3020 gram of the amide gave .1472 gram of water.

IV. 3003 gram of the amide gave .1450 gram of water.

V. .2391 gram of the amide gave .3502 gram of barium sulphate.

VI. .2910 gram of the amide gave .3950 gram of barium sulphate.

VII. .2210 gram of the amide gave .0202 gram of nitrogen.

VIII. .1800 gram of the amide gave .0149 gram of nitrogen.

IX. .2185 gram of the amide gave .0173 gram of nitrogen.

The sulphur determinations were made by Liebig's method. Nitrogen determination VII was by the Dumas method; VIII and IX were by Wilfarth's modification of the Kjeldahl method, and were made by Mr. J.W.Lawson of this laboratory, who also made the carbon and hydrogen determinations.

	Calculated for $C_6H_4\genfrac{}{}{0pt}{}{CH_3}{SO_2NH_2}$		Found.	
		I.	II.	III.
C	49.12	49.17	49.06	
H	5.26	5.40	5.36	
S	18.71	18.92	18.73	
N	8.18	8.13	8.28	8.15

The amide obtained by Müller, was described as crystallizing at 96° in slender needles. Though a great many observations were made, in no case did the amide come down in this form, hexagonal scales sometimes bunched in fern-shaped leaves, being the rule without exception.

Pechmann's amide came down in rhombic scales, and needles, and though it melted at 104°, was far from being pure meta-toluene-sulphon-amide, for, as evidenced by Metcalf's work, it contained a large quantity of para-ethoxy-meta-toluene-sulphon-amide. Pagel's amide differed from that herein described in being tolerably easily soluble in cold water and in crystallizing from all solvents in large leaves. The other amides referred to above, which were prepared after the methods of Müller and Pechmann, or a combination of both, can not be considered as pure products.

Decomposition of Meta-toluene-sulphon-amide. — As a test of the purity of the amide obtained from the product of the diazo decomposition in presence of alkalies and other substances, the experiment by which Otto[1] showed that Beckurts' amide was a mixture of the para and ortho compounds was performed.

15 grams of the amide were heated to $150°$ with concentrated hydrochloric acid, in a sealed tube, for four hours. The product of the reaction was very easily soluble in cold water, and this solution was repeatedly boiled down with water till free of hydrochloric acid. It was then treated with potassium hydroxide solution until all the ammonia was driven off. The potassium salt thus formed was very easily soluble in hot 95% alcohol, from which it crystallized out on cooling in a mass of fine needles. These crystals were dried in an air-bath at a temperature of $140°$, powdered, and treated with an equal weight of phosphorus pentachloride. The reaction was quiet and gentle, and a quantity of colorless oil was the result. Poured into cold water, a white scum covered the surface of the oil, but this disappeared on

---

[1] Berichte 13 - 1292.

standing. The oil was separated from the water, poured into a test-tube and placed in a freezing-mixture for half an hour. At the expiration of this period it seemed to be as limpid as in the beginning, and showed no trace of a solid formation.

OXIDATION OF META-TOLUENE-SULPHON-AMIDE.

Another experiment was made, having for its aim the oxidation of meta-toluene-sulphon-amide to meta-sulphamine-benzoic acid. To a mixture of 80 grams of concentrated sulphuric acid diluted with three times its volume of water, and 60 grams of potassium bichromate, 20 grams of meta-toluene-sulphon-amide were added. The flask containing the mixture was connected with a reflux-condenser and heated gently over a low flame. The action began almost immediately, and became so energetic that cold water had to be added to the contents to prevent foaming over. After boiling for an hour, the oily layer of melted amide which covered the surface entirely disappeared, and the flask was removed and its contents, after cooling, were poured into cold water. A mass of crystals separated out. These were drained by

means of a filter-pump, and washed with water till free from all bichromate color. They were then treated on the funnel with cold sodium-carbonate solution, when most of the acid dissolved. The substance which remained on the filter was examined, and it proved to be pure meta-toluene-sulphon-amide. It weighed about 2 grams.

META-SULPH-AMINE-BENZOIC ACID.

The sodium salt which passed in solution through the filter, was treated with hydrochloric acid and a quantity of minute scales separated out. These were filtered off, dried, and recrystallized. They gave a melting-point of $255°$ (uncorrected). From the 20 grams of meta-toluene-sulphon-amide, a yield of 12 grams of meta-sulphamine-benzoic resulted. It was carefully compared with the meta-sulphamine-benzoic acid obtained by Limpricht and Uslar,[1] from sulpho-benzamide and potassium hydroxide, which was described as separating from the cooled solution in scales which melted above $200°$. This acid was sparingly soluble in cold water, more easily in ether, and very soluble in alcohol.

[1] Liebig's Annalen, 103 - 35.

Its barium salt was made by adding barium carbonate to a boiling solution of the acid. This was easily soluble in water, from which it crystallized out in small masses possessing a "wavellite"[1] structure, and containing four molecules of water of crystallization. Its silver salt consisted of long silky needles which contained two molecules of water of crystallization and did not blacken on boiling or long exposure to the light.

A small portion of the meta-sulphamine-benzoic acid made by oxidizing meta-toluene-sulphon-amide, was gently heated in a test-tube in a sulphuric acid bath to $245° - 250°$ for 45 minutes. Upon cooling, it stiffened to a brown, glassy mass, which was easily soluble in cold water, from which it crystallized out, on evaporating down the solution, in long needles.

Barium Salt of Meta-sulphamine-benzoic acid. - The barium salt of the acid was made by neutralizing a boiling solution of the acid by means of pure barium carbonate. The undecomposed carbonate was filtered off and the solu-

[1] "Wawellitartigen."

tion evaporated down. Upon standing, small radiating nodules formed, which corresponded to the "wavellite" character of Limpricht and Uslar's barium salt.

The analysis of the barium salt showed its composition to be $Ba(C_6H_3 <^{CO_2}_{SO_2NH_2})_2 + 4\frac{1}{2} H_2O$.

I. .2345 gram of the barium salt, dried by exposure to the air, lost, upon being heated to 180°, .0305 gram water.

II. .3022 gram of the salt, gave off at 180°, .0394 gram of water.

III. .2039 gram of the anhydrous barium salt, gave upon ignition with sulphuric acid, .0880 gram of barium sulphate.

IV. .2623 gram of the anhydrous salt, gave .1126 gram of barium sulphate.

	Calculated.	Found.
H_2O	13.10	13.05 - 13.03
Ba	25.51	25.37 - 25.24

The dehydrated salt takes up water very rapidly, which makes its weighing difficult. The difference in results for the water of crystallization of this and Limpricht

and Uslar's determination may be due to this, and also to the fact that their barium salt was dried over sulphuric acid, and was heated but to 110°.

<u>Silver Salt.</u> - The silver salt was made by boiling a solution of the acid with silver carbonate. It was soluble to a slight extent in boiling, and insoluble in cold water. From hot water it crystallized out in long, fine needles, which slowly blackened upon exposure to the light. It contained no water of crystallization, and analysis showed that it corresponded to the formula $C_6H_4\begin{smallmatrix}COOAg\\SO_2NH_2\end{smallmatrix}$

I. .2256 gram of the salt gave .0779 gram silver.

II. .2710 gram of the salt gave .0945 gram silver.

	Calculated.	Found.
Ag	35.06	34.53 - 34.90

Limpricht and Uslar ascribed two molecules of water of crystallization to this salt.

On the whole, if we except the water of crystallization of the silver salt, there are enough points of similarity between the salts of meta-sulphamine-benzoic acid obtained by the oxidation of meta-toluene-sulphon-amide, and those

ty of both substances.

ON META-TOLUENE-SULPHONIC ACID.

INTRODUCTION.

The action of sulphuric acid on toluene was first investigated by Jaworsky,[1] who described one toluene-sulphonic acid as the resulting compound.

In 1869, Engelhardt and Latschinoff[2] heated toluene with sulphuric acid, and made the potassium salt of the resulting acid. This crystallized out in two forms; prisms which were difficultly soluble, and needles which were more easily soluble in water. These two potassium salts were fused with caustic potash and gave two isomeric cresols, the para and ortho compounds.

Barth,[3] working in a similar manner, obtained by fusing

---

[1] Zeitschrift für Chemie, I - 272.   [2] Ibid, 1869 - 517.
[3] Liebig's Annalen, 152 - 91.

the crude mixture of potassium salts, para-oxybenzoic and salicylic acids.

Anna Wolkow[1] confirmed this work, but ascribed the meta position to the sulphon group in the acid which passed over into salicylic acid, and further, made two sulphon-chlorides, one liquid and one solid, and two amides, that from the α or para acid melting at 137°, and that from the β or, as it was then called, the meta acid melting at 153°—154°.[3]

Fittig and Ramsay,[2] in order to settle definitely the position of the sulphon group in the acid which Anna Wolkow called the meta-toluene-sulphonic acid, proceeded as follows: They treated toluene with sulphuric acid, made the isomeric potassium salts, and separated as far as possible by fractional crystallization, the para salt. The mother liquid, which contained some of the para and an isomeric salt which was either ortho or meta, was then evaporated to

[1] Zeitschrift fur Chemie, 1870-391. [2] Liebig's Annalen,138-242.

[3] This was previous to V.Meyer's work on the orientation of the hydroxy benzoic acids.

dryness and thoroughly mixed with potassium cyanide. This
mixture was distilled in an iron retort; the oily liquid
which came over was heated with caustic potash till free
from ammonia. After washing with ether, the potassium salt
was acidified and the toluic acid extracted with ether.
This was a mixture of para-toluic with either ortho- or
meta-toluic acid. An easy method of separation was found
by treating the calcium salts with alcohol in which the
para salt was not soluble. By this means they obtained, not
the meta, but the ortho-toluic acid. They further confirmed their work by heating the acid thus obtained with a mixture of potassium bichromate and dilute sulphuric acid, and
obtained as a product neither isophthalic nor terephthalic
acid, which showed conclusively that the isomeric acid
formed together with para-toluene-sulphonic acid, when toluene is treated with sulphuric acid is not the meta, but
the para compound.

The next contribution to our knowledge of the isomeric-
toluene-sulphonic acids, was a consequence of Hübner and
Post's[1] work on bromo-toluene and the behavior of its hydro-

---
[1] Liebig's Annalen, 169 - 47.

drogen atoms.

By the action of sulphuric acid on crystallized bromotoluene, which they showed to be a para compound, two isomeric bromo-toluene-sulphonic acids were obtained, which they designated as the α and β para-bromo-toluene-sulphonic acids. The β acid was treated with sodium amalgam in order to remove the bromine atom, and the toluene-sulphonic acid thus obtained, was fused with potassium hydroxide. Salicylic acid was obtained, which was evidence that the sulphon group was in the position ortho to the methyl group. It was, therefore, by exclusion, concluded that the α acid must have the sulphon group in the meta position.

F.C.G.Muller[1] made the barium salt of ortho-brom-toluene-sulphonic acid and treated it in solution, with sodium amalgam in order to remove the bromine atom from the molecule. The excess of alkali was neutralized with sulphuric acid, the sodium sulphate removed by evaporation and crystallization, and the residue dried. The dried mass was treated with phosphorus pentachloride, and an oily toluene-sulphon-chloride was obtained. This was heated with water to

[1] Liebig's Annalen, 169 - 47.

130°, and the resulting acid mixture was freed from hydrochloric acid by evaporation and the passage of a current of air through it. The toluene-sulphonic acid was obtained as a syrup which stiffened to a crystalline condition. Various salts of the acid were made, some of which will be described later on. The amide obtained melted at 90°—91°. Because this toluene sulphonic acid differed from both the ortho and para acids, Müller considered it to be meta-toluene-sulphonic acid.

About the same time, F.Cervor[1] made ortho-diazo-toluene-sulphonic acid by subjecting ortho-toluidine-sulphonic acid to the action of nitrous fumes. The diazo compound was decomposed in alcohol under pressure. The sodium salt of the sulphonic acid obtained was made, and gave, upon treatment with phosphorus pentachloride, a fluid sulphon-chloride, which was converted by the action of ammonia into an amide melting at 146°, and differing in this and other properties, from the ortho and para amides previously known.

In 1874, Pechmann[2] obtained a quantity of para-toluidine-meta-sulphonic acid by heating sulphuric acid with para-tol-

[1] Liebig's Annalen, 159-333. [2] Ibid, 173-195.

uidine, and crystallizing out the para-toluidine-ortho-sulphonic and the di-sulphonic acids formed, leaving the para-toluidine-meta-sulphonic acid, on account of its greater solubility, in solution.

From this solution the para-toluidine-meta-sulphonic acid crystallized out in sulphur-yellow needles which could not be decolorized by repeated boiling with animal charcoal. By precipitation from its salts, however, it was obtained colorless. It was difficultly soluble in cold water, one part of acid requiring ten parts of water for its solution.

The diazo- compound was made by suspending the finely divided acid in alcohol, and treating it with nitrous fumes. This was then decomposed, by heating it under pressure with alcohol.

Various salts of the resulting sulphonic acid were made which showed some difference from that of Muller, also an amide which melted somewhere below $100°$.

In 1875, Pagel,[1] working on ortho-toluidine-sulphonic acid, obtained a toluene sulphonic acid which resembled

[1] Liebig's Annalen, 176 - 297.

Müller's meta acid.

A mixture of alcohol and sulphuric acid was neutralized with lime, and the calcium-ethyl sulphate was decomposed by the addition of a solution of the ortho-toluidine salt of oxalic acid. By filtration and crystallization, crystals of ortho-toluidine-ethyl-sulphate were obtained, which were decomposed into alcohol and ortho-toluidine-sulphuric acid by being heated to $180°-200°$ in an air bath.

The ortho-toluidine-sulphonic acid was changed into the diazo compound, and this transformed into bromo-toluene-sulphonic acid.

By proceeding as Müller did, a toluene-sulphonic acid was obtained which agreed tolerably well with Müller's meta-toluene-sulphonic acid, if we except the melting-point of the amide, which Papel found to be $104°$.

In 1877, H. Beckurts[1], seeking for a method by which to obtain a good yield of pure ortho-toluene-sulphonic acid, heated a mixture of ordinary sulphuric acid and toluene and

[1] Berichte X - 945.

made the potassium salts of the resulting sulphonic acids in the usual manner. By treatment with phosphorus pentachloride, a mixture of sulphon-chlorides was obtained, from which the para-toluene-sulphon-chloride crystallized out, upon cooling the mixture below zero.

The remaining liquid sulphon-chloride was transformed into the amide compound, which was purified and subjected to fractional crystallization. Two compounds were obtained, one difficultly soluble in alcohol, which melted at 153° - 154°, the other easily soluble in the same medium, and melting at 104°. From this work, Beckurts concluded that the action of sulphuric acid on toluene, resulted in the formation of the three isomeric acids. The work was pursued further, and some of the amide heated with hydrochloric acid in a sealed tube to 150°. The ammonia salt of the acid was thus obtained, and from a portion of this, treated with barium hydroxide, the barium salt. From a comparison of these salts with the ammonium and barium salts obtained by Müller, Beckurts asserted that they were salts of meta-toluene-sulphonic acid.

Fahlberg,[1] working in this laboratory, doubted the fact that Beckurts had obtained meta-toluene-sulphonic acid because, though his product showed differences from the ortho and para compounds, he neglected to prove its constitution by converting it into an oxybenzoic acid. Fahlberg repeated Beckurts work, with the sole exception that water, and not alcohol, was used as the medium for fractional crystallization.

The liquid sulphon-chloride, from which no more solid para-sulphon-chloride separated out when cooled to 15° below zero, was treated with aqueous ammonia. As products there were obtained α toluene-di-sulphon-amide, melting at 186°–187°, and a yellow mixture of amides. This latter he subjected to fractional solution; whereby only a portion dissolved. The undissolved amide melted at 153° – 154°, and gave upon fusing with caustic potash, salicylic acid. Therefore it was ortho-toluene-sulphon-amide. From the soluble portion of the mixture, two compounds were obtained by fractional crystallization. That less soluble, on puri-

[1] American Chemical Journal,1 - 170, -- Berichte,XII,-1043.

fication, proved also to be ortho-toluene-sulphon-amide..

From the solution of the more soluble portion an amide crystallized out, which melted at 120°, but repeated purifications with animal charcoal lowered the melting-point to 108°, where it remained constant. A quantity of the amide which melted at 108° was oxidized by means of potassium permanganate, and three products were obtained; para-sulphamine-benzoic acid, anhydro-ortho-sulphamine-benzoic acid, and acid potassium-ortho-sulpho-benzoate; which were the products one would expect to result from the oxidation of a mixture of para-and ortho-toluene-sulphon-amides.

Further evidence that Beckurts' amide, despite the constancy of its melting-point, was a mixture, was given by the following experiments: Equal parts of pure para and ortho-toluene-sulphon-amides were dissolved in water and the mixture subjected to crystallization. Two forms of crystals were obtained: long fine needles which melted constantly at 120°, and leaves which fused at 108°. Slight impurities lowered the melting-point to 104° — 105°, as in the case of Beckurts' amide.

F.H.S. Müller,[1] in 1879, mentioned the formation of meta toluene-sulphonic acid by the action of sulphurous acid on the corresponding diazo compounds; but beyond the melting-points of the amide, anilide and toluidide, gave no details of his work.

Claesson and Wallin,[2] in 1879, claimed that they had obtained meta-toluene-sulphonic acid by the action of chlor-sulphonic acid on toluene. These investigators worked with large quantities of material bringing together, in small portions at a time, 2800 grams of chlorsulphonic acid and 1119 grams of toluene.

The solid para-sulphon-chloride was removed by cooling the mixture, and the liquid chlorides converted into the amides. Two amides were separated by fractional crystallization, the ortho compound melting at $153°-154°$, and the other, supposed to be the meta, melting constantly at $107°-108°$. This latter was transformed into the toluene sulphonic acid by being heated with water in a sealed tube to $150°$, and various salts were made. These were compared with the salts made by Müller, and also with the analogous

[1] Berichte XII - 1348. [2] Ibid XII - 1848.

salts of the para and ortho acids. From this, despite the fact that they had knowledge of Fahlberg's work, Claesson and Wallin concluded that all three isomeric acids were formed as a result of sulphuric acid on toluene.

The next step in the controversy was taken by R.Otto[1] who, at the request of Beckurts, examined carefully the specimen of so-called meta-toluene-sulphon-amide which the latter had obtained three years previously, and which was claimed by Fahlberg to be a mixture, and by Claesson and Wallin, to be meta-toluene-sulphon-amide. The substance, weighing about ten grams, was heated with hydrochloric acid in a sealed tube. The sulphonic acid formed was converted into the potassium salt, and this was treated with phosphorus pentachloride, with the formation of an oil which soon stiffened and the solid portion of which was freed from the fluid by pressure. This latter portion dissolved in benzene, from which it crystallized out in large scales which had the same melting-point as para-toluene-sulphon-chloride, and gave a corresponding amide and anilide. This was conclusive evidence that meta-toluene-sulphonic acid

---

[1] Berichte XIII - 1292.

was not obtained by Peckurts.

Neville and Winther,[1] in 1880, investigating the formation of amido-sulphonic acids by the action of concentrated sulphuric acid, obtained a toluene sulphonic acid by heating ortho-diazo-toluene-sulphonic acid under pressure with alcohol, and by reducing bromo-toluene-ortho-sulphonic acid by means of sodium amalgam. By treating the potassium salt with phosphorus pentachloride, they obtained a liquid sulphon-chloride which gave, in both cases, an amide melting at $106.5° - 107.5°$, whence they concluded that the acid which they had obtained was meta-toluene-sulphonic acid.

In 1886, Vallin,[2] convinced by the work of Otto, that Fahlberg's views were correct, and that he and Klason had not obtained meta-toluene-sulphonic acid, again attacked the problem of obtaining the compound. Pechmann's method was followed; an acid was obtained, and a series of salts made and described.

---

[1] Journal of the Chemical Society, 37-398.   [2] Berichte, 19-2952.

The work of Metcalf[1] furnished evidence that the decomposition of para-diazo-meta-toluene-sulphonic acid by boiling with alcohol under pressure, resulted in the formation of para-ethoxy-meta-toluene-sulphonic acid, in far greater proportion than meta-toluene-sulphonic acid. Consequently, neither by Pechmann nor by Wallin had the acid been obtained

Preparation of Meta-Toluene-Sulphonic Acid. - In consideration of all this dissentient testimony in regard to meta-toluene-sulphonic acid, and the availibility of pure meta-toluene-sulphon —amide in any desired quantity, it was thought desirable to transform the amide into the acid and study its properties. Sealed tubes containing 25 grams of the pure amide, and a quantity of hydrochloric acid were heated to 140° for six hours. Upon cooling, the tubes contained a mass of clear white fern-shaped crystals, entirely different in form from the amide crystals which had been introduced therein. The tubes were opened, and their contents were poured upon a filter, and freed from liquid by

[1] American Chemical Journal, 15-301 and Dissertation.

suction. The crystals were dried and examined. They proved to consist entirely of ammonium chloride.

The fluid portion of the contents of the tube was evaporated down, on a water-bath, water added, and reevaporated, and this process repeated until the greater portion of the hydrochloric acid was removed. It was then boiled with potassium hydroxide solution until no further evolution of ammonia was evident. The solution, which contained the potassium salt of meta-toluene-sulphonic acid and some potassium chloride, was then evaporated down until it solidified in a mass of scales. These were powdered and boiled with alcohol of 95%. The greater portion of the salt dissolved, and separated out on cooling, into a mass of minute scales which nearly filled the beaker. From the mother-liquor more were obtained by evaporation. The undissolved portion was subjected to several more extractions until potassium meta-toluene-sulphonate no longer crystallized out of the solution. The residue was then examined and found to be potassium chloride. The crystals from the alcoholic solution were collected on a filter and dried. They were then powdered, and heated in an air-bath for an hour to 150°.

Upon cooling, the substance was rubbed up in a mortar with an equal weight of phosphorus pentachloride. The action was immediate and brisk, and the contents of the mortar became liquid; but the decomposition was not nearly as violent as when the potassium salt obtained from the diazo decomposition was employed. The liquid chloride mixture was poured into water, and thoroughly stirred up. A clear, almost colorless oil gathered on the bottom of the vessel. This was washed several times with water, and then placed in a test-tube in a freezing-mixture. It did not solidify. Müller converted the sulphon-chloride, which he obtained, into the corresponding acid by heating it with water in a sealed tube. He stated that it did not decompose with water until it was heated to $130°$.

About 5 cc. of the sulphon-chloride and 100 cc. of water were placed in a flask connected with an inverted condenser and boiled. Considerable bumping occurred, a small portion of the contents of the flask at times fuming out through the top of the condenser. In about an hour and a half, the oily liquid had entirely disappeared. The solution was then poured into an evaporating dish and boiled

until it attained a syrupy consistency. More water was added and it was again evaporated down. This process was repeated until the solution gave no indication of the presence of hydrochloric acid when tested with silver nitrate. All efforts to obtain the meta-toluene-sulphonic acid in crystalline condition, were unsuccessful. After it had become viscous, further heating on a water-bath resulted in giving it a dark, almost black color. Made clear by boiling with animal charcoal and filtration, it darkened again when kept on the water-bath.

Efforts were then made to crystallize it by evaporation under diminished pressure. About 15 cc. of the syrupy liquid was placed in a round-bottomed flask, which was connected to a filter-pump and kept in boiling water for over an hour. Some water distilled over in the beginning, but ceased coming over after a while. On cooling, the contents of the flask remained fluid. The viscous liquid was then placed in a test-tube and kept in a freezing-mixture for half an hour, and though its temperature was lowered to $-18°$, its condition did not change.

Müller, Pechmann, Pagel and Klason, all obtained, by

evaporating down their acid solutions, a syrup which stiffened to a mass of deliquiscent crystals.

Barium salt of meta-toluene-sulphonic acid. — A solution of the acid was heated to boiling, and neutralized with barium carbonate. The solution was then filtered and evaporated down to a small bulk. Upon cooling, the barium salt crystallized out in minute scales. With favorable conditions, bunches of apparently rectangular plates, about 1/8 of an inch long, were obtained. It is very soluble in water, insoluble in absolute alcohol and ether; soluble in dilute alcohol, and precipitated from its solution in the latter by the addition of a large quantity of ether.

It contains one molecule of water of crystallization.

The salt was analyzed with the following results:

For water of crystallization:

I. .2702 gram of the salt gave off, at 180°, .0101 gram of water.

II. .2208 gram of the salt gave off, at 180°, .0080 gram of water.

For barium:

III. .2702 gram of the salt gave .1260 gram of barium sulphate.

IV. .2208 gram of the salt gave .1030 gram of barium sulphate.

	Calculated for $Ba(C_6H_4{<}^{CH_3}_{SO})_2 + H_2O$	Found. I.	II
H_2O	3.62	3.73	3.62
Ba	27.56	27.42	27.42

The barium salt made by Müller by boiling the free acid with pure barium carbonate, gave off two molecules of water of crystallization. Neither from water nor from alcohol could he obtain it, in a distinctly crystalline condition.

Lead Salt.— Pure lead carbonate was decomposed by boiling with a small quantity of the dilute acid. The solution, evaporated down and left to stand, filled up with a mass of crystals consisting of small scales. This salt was very soluble in water, and slightly soluble in strong alcohol. The addition of a large quantity of ether to its dilute alcoholic solution precipitated the lead salt in

scales. The analysis of the salt was made with the following results:

For water of crystallization:

　　I.　.2480 gram of salt, heated to 160°, gave off .0078 gram of water.

　　II.　.2394 gram of salt gave off .0078 gram of water.

　　III.　.2212 gram of salt gave off .0070 gram of water.

For lead:

　　IV.　.2480 gram of the anhydrous salt, gave .1278 gram of lead sulphate.

　　V.　.3538 gram of the salt gave .1813 gram of lead sulphate.

　　IV.　.3781 gram of the salt gave .1923 gram of lead sulphate.

Calculated for $Pb(C_6H_4\genfrac{}{}{0pt}{}{CH_3}{SO_3})_2 + H_2O$　　　Found

		I.	II.	III.
H_2O	3.17	3.14 -	3.17 -	3.16
Pb	36.33	35.21 -	35.57 -	34.74

In describing his lead salt, Müller calls attention to the fact that the anhydrous salt absorbed moisture very rapidly, and made it difficult to get good results for the

water of crystallization. No such behavior was noticed in
the case of the lead salt here described. The results for
water of crystallization agreed as closely as possible with
the theoretical, and no difficulty in weighing was experi-
enced. On the other hand, it was found impossible to get
good results in the determination of lead. Many analyses
were made, the three given above, being the best results
obtained. Ignition of the salt, alone, or when moistened
with sulphuric or nitric acid, invariably caused some reduction. Precipitation in alcoholic solution, as lead sulphate, and filtration through a Gooch crucible, gave no
better results.

Calcium Salt. - The calcium salt was made by decomposing finely powdered calcite by means of the free acid.
It was extremely soluble in water, and could be obtained in
crystalline condition only by evaporating the solution almost to dryness, when, upon cooling, it solidified in a
mass of scales, which were so matted together, that they
could not be obtained in a dry condition, by long standing
in the air. Upon heating a saturated solution of the cal-

cium salt, a slight milky precipitation occurred, which disappeared on standing, indicating its greater solubility in cold than in hot water. Analyses were made of a specimen which had been dried by standing over sulphuric acid.

I. .3969 gram of the calcium salt, heated to $200°$, lost .0343 gram of water.

II. .3196 gram of the calcium salt, heated to $200°$, lost .0270 gram of water.

III. .3969 gram of the salt gave .1310 gram of calcium sulphate.

IV. .3196 gram of the salt gave .1051 gram of calcium sulphate.

Calculated for $Ca(C_6H_4\langle{}^{CH_3}_{SO_3})_2 + 2H_2O$.   Found.

		I.	II.
H_2O	8.61	8.59	8.44
Ca	9.57	9.66	9.67

Müller did not succeed in obtaining the calcium salt of his toluene-sulphonic acid in a crystalline condition from its solution in water, and his salts, whether obtained by evaporating the water solution to dryness or by filtra-

tion of the boiling alcoholic solution through a warmed filter, continued to give off water when heated to $200°$.

The salt, the analysis of which is given above, was easily obtained in crystalline condition, and gave off all its water of crystallization at $180° - 190°$; and heated to $200°$, it began to darken in color. It was evidently a different compound from that of Müller.

Sodium Salt - The sodium salt was made by exactly precipitating the barium by means of sodium sulphate, from a solution of the barium salt. It is very soluble in water, from which it crystallizes out in large scales when the solution is evaporated down.

Its analysis resulted as follows:

For water of crystallization:

I. .1641 gram of salt lost, upon heating to $150°$, .0139 gram of water.

II. .2030 gram of salt gave off .0175 gram of water.

For sodium:

III. .1502 gram of the salt gave .0549 gram of sodium sulphate.

IV. .1885 gram of the salt gave .0533 gram of sodium sulphate.

Calculated for $NaC_7H_4 \begin{cases} CH_3 \\ SO_3 \end{cases} + H_2O$

		Found.	
		I.	II.
H_2O	8.49	8.47	8.49
Na	10.85	10.91	10.73

The sodium salt obtained by Müller did not crystallize well from water, and contained but one half a molecule of water of crystallization.

Potassium Salt. – The potassium salt was made, as was the sodium, by the exact precipitation of a solution of the barium salt. This salt crystallized in thin scales, and was easily soluble in water, sparingly so in strong alcohol. It dissolved to a great extent in hot, ordinary alcohol, from which it separated out in small scales. Addition of ether to the alcoholic mother-liquor caused a copious precipitation.

The following are the results of its analysis:

I. .2143 gram of the salt, heated to 180°, gave off .0096 gram of water.

II. .1980 gram of the salt, gave off .0089 gram of water.

III. .2146 gram of the salt gave .0844 gram of potassium sulphate.

IV. .1980 gram of the salt gave .0773 gram of potassium sulphate.

Calculated for $K\ C_6H_4{<}{}^{CH_3}_{SO_3} + \frac{1}{2} H_2O$.   Found.

		I.	II.
H_2O	4.11	4.47	4.49
K	17.75	17.63	17.50

This potassium salt agrees with that of Müller in solubility and water of crystallization.

Silver Salt - Pure silver carbonate, obtained by precipitating a solution of silver nitrate by means of sodium carbonate, was decomposed in a boiling solution of the acid. Upon filtering and evaporating down to a small volume, bunches of rectangular plates were obtained, which showed a micaceous cleavage on breaking apart. These were easily soluble in water, and to a much lesser extent in alcohol. From their solution in the latter, they were precipitated by

the addition of ether. They contained no water of crystallization, and blackened on being exposed to the light.

Analyzed:

I. .2061 gram of the salt gave .0080 gram of silver.
II. .2640 gram of the salt gave .1023 gram of silver.

Calculated for $C_6H_4{<}^{CH_3}_{SO_3Ag}$ Found.

		I.	II.
Ag	38.71	38.31	38.75

<u>Magnesium Salt</u>. — The magnesium salt was made by boiling pure magnesium carbonate in a solution of the acid. After neutralizing the acid, the solution was filtered and reduced in volume by evaporation. On standing, the salt crystallized out in beautiful long transparent thick prisms which lost water rapidly in the air. Their crystalline structure appeared to be orthorhombic and hemihedral. The magnesium salt is slightly soluble in alcohol, from which it is precipitated by ether, in minute plates. The property of rapidly losing water of crystallization when exposed to the air, was not noticed in any of the other salts of meta-

toluene-sulphonic acid.

Subjected to analysis it yielded the following results: For water of crystallization.

I. .2050 gram of the salt lost, when heated to 150°, .0644 gram of water.

II. .2166 gram of the salt gave off, at the same temperature, .0691 gram of water.

Calculated for $Mg(C_6H_4{<}^{CH_3}_{SO_3})_2 + 10\ H_2O$. Found.

		I.	II.
H_2O	32.05	31.45	31.90

For magnesium:

I. .2636 gram of the anhydrous salt gave .0865 gram of magnesium sulphate.

II. .2797 gram of the anhydrous salt gave .0896 gram of magnesium sulphate.

Calculated for $Mg(C_6H_4{<}^{CH_3}_{SO_3})_2$ Found.

		I.	II.
Mg	6.56	6.55	6.40

<u>Manganese Salt.</u> - A solution of the barium salt was treated to exact precipitation with manganous sulphate.

Upon reducing the solution to a small bulk and cooling, the manganese salt crystallized out in long prisms, which were easily soluble in water, and to a slight extent in alcohol. From the alcoholic solution they were precipitated in the form of bunches of small needles.

The following are the results of the analyses:

I. .2150 gram of the salt gave off, when heated to $200°$, .0461 gram of water.

II. .2708 gram of the salt gave off .0588 gram of water.

III. .2150 gram of the anhydrous salt gave .0326 gram of manganous manganic oxide..

IV. .1589 gram of the salt gave .0326 gram of manganous manganic oxide.

Calculated for $Mn(C_6H_4{<}^{CH_3}_{SO_3})_2 + 6H_2O$ Found.

		I.	II.
H_2O	21.39	21.44	21.67
Mn	10.89	10.92	10.93

<u>Zinc Salt.</u> - The zinc salt was made by treating a boil-

ing solution of the acid with pure zinc carbonate. After filtering and evaporating down the solution, long slender prisms were deposited. They were soluble in water and alcohol, and were precipitated from the alcoholic solution in minute scales upon the addition of ether. The salt was very stable, and the crystals retained their lustre after long exposure to the air. It was analyzed for zinc, by precipitating it as carbonate and igniting the latter.

I. .3377 gram of the salt, lost upon being heated to 180°, .0695 gram of water.

II. .3106 gram of the salt lost, upon being heated to 180°, .0638 gram of water.

III. .3377 gram of the salt gave .0527 gram of zinc oxide.

IV. .3106 gram of the salt gave .0482 gram of zinc oxide.

Calculated for $Zn(C_6H_4{<}{}^{CH_3}_{SO_3})_2 + 6 H_2O$ Found.

		I.	II.
H_2O	20.97	20.58	20.54
Zn	12.62	12.51	12.52

Copper Salt. - The copper salt was obtained by precipitating barium sulphate from a solution of the barium salt by means of copper sulphate. Upon evaporating the solution nearly to dryness and cooling, a mass of light blue scales was obtained. The copper salt was very soluble in water and ordinary alcohol, very slightly soluble in absolute alcohol, and insoluble in ether. The addition of ether to an alcoholic solution of the salt, caused its precipitation in minute, almost white, scales. No very satisfactory determination of water of crystallization of the copper salt could be obtained.

On heating .2995 gram of the salt to constant weight in an air-bath at $170°$, it was found to have lost .0450 gram of water, or 15.02 %. The theoretical percentage of water, assuming the salt to contain four molecules of water of crystallization, is 15.03. The determination of the water of crystallization of different specimens obtained by precipitating the salt from its alcoholic solution by means of ether, which varied in color from almost white to light blue, gave no concordant results.

The copper was determined in the anhydrous salt by the

electrolytic method.

I. .2545 gram of the anhydrous salt gave .0401 gram of copper.

II. .3725 gram of the salt gave .0585 gram of copper.

Calculated for $Cu(C_6H_4{<}{}^{CH_3}_{SO_3})_2$ Found.

		I.	II.
Cu	15.55	15.70	15.70

<u>Meta-toluene-sulphon-anilide.</u> - This substance was made by bringing together some meta-toluene-sulphon-chloride and a slight excess of aniline. The action was extremely rapid. The flask became very hot and its contents solidified in a crystalline mass. The contents of the flask were then boiled with dilute hydrochloric acid, in order to remove any unacted upon aniline, and the anilide was then crystallized from ether, in which it was very soluble. A yield of small prismatic crystals was obtained. The anilide was insoluble in cold water and dissolved only to a very slight extent in boiling water. It was extremely soluble in alcohol, from which it crystallized out in large monoclinic prisms. It melted sharply at 96°.

Fuller and Wiesinger[1] mention a meta-toluene-sulphon-anilide, which melted at 72°. They obtained the meta acid by treating the diazo compound with sulphurous acid.

Meta-toluene-sulphon-toluide. — The ortho- and para toluides of meta toluene sulphonic acid were made by bringing together meta-toluene-sulphon-chloride and the proper toluidine, the latter in slight excess. The reaction was very rapid and violent, part of the contents of the flask volatilizing. The excess of toluidine was removed by warming the mixture with dilute hydrochloric acid, and washing several times with warm water. Both toluides are well-crystallized compounds, insoluble in water, but very soluble in ether, alcohol and benzene.

The meta-toluene-sulphon-ortho-toluide crystallized in prisms and thick plates, and melted at 108°.

The meta-toluene-sulphon-para-toluide crystallized in small prisms and melted at 105°. The toluide mentioned by Müller and Wiesinger[2], melted at 103°.

---

[1] Berichte XII - 1348.   [2] Ibid XII - 1348.

CONCLUSION.

From the experiments detailed in the first part of this work, we may conclude that:

1. When para-diazo-meta-toluene-sulphonic acid is decomposed in alcohol in the presence of an excess of some alkalies and zinc dust, only the hydrogen reaction takes place.

2. When the decomposition takes place in alcohol saturated with ammonia, the ammonium salt of para-toluidine-meta-sulphonic acid is formed.

3. The nature of the alcohol, as regards methyl and ethyl alcohol, has no influence on the reaction.

4. If we except the case of decomposition in the presence of calcium carbonate, countenance is given to the suggestion that the neutralization of free acid by the presence of some substance which can combine with it, determines

the hydrogen reaction.

But why calcium carbonate does not effect this purpose, which, on the other hand, is easily accomplished by zinc dust, has yet to be determined.

From the study of meta-toluene-sulphonic acid and its salts, we may conclude that the compounds obtained by Müller and others who employed analogous methods for obtaining them, were impure products.

This follows especially from a comparison of the properties of the amide, and from the fact that the sulphon-chloride suffers decomposition into the acid when boiled with water in an open vessel, while the sulphon-chloride of Müller did not decompose until heated to 150° in a sealed tube.

No detailed comparison of the salts of meta-toluene-sulphonic acid herein described, with those made by other methods than those of Müller was made, for previous work in this laboratory, by Fahlberg and Metcalf, showed conclusively that they were not true meta compounds.

The temperatures given in the foregoing work are all uncorrected.

The analyses were calculated on the basis of the following atomic weights:

H	1	Na	23
O	16	K	39
C	12	Ca	40
N	14	Ba	137
Ag	103	Cu	63
Mg	24	Mn	55
Zn	65	Pb	207
S	32		

Johns Hopkins University, Baltimore, Md., May, 1895.

BIOGRAPHICAL SKETCH.

John Joseph Griffin was born near Corning, New York, June 24, 1859. His early education was obtained in the public schools of Lawrence, Mass., graduating from which he entered Ottawa (Canada) College in 1878. He received the degree of A.B. from this institution in 1881, and two years later, that of A.M., and in the same year entered the Ottawa Diocesan Seminary. Ordained priest, May 1, 1885, he spent a year as instructor in elementary physics in Ottawa College, after which he undertook the work of the ministry in the archdiocese of Boston. In September, 1887, he returned to Ottawa College as instructor in elementary physics and chemistry, which position he held for three years. In 1890, he entered Johns Hopkins University as a graduate student in chemistry, with physics and mathematics as subordinate subjects. While pursuing his studies at Johns Hopkins University, he conducted classes in chemistry at St. Joseph's Seminary and Notre Dame of Maryland.

www.ingramcontent.com/pod-product-compliance
Lightning Source LLC
Chambersburg PA
CBHW031449160426
43195CB00010BB/913